Valsergio Barros da Silva
Janiquelle da Silva Rabelo
Antonio V. Rodrigues de Almeida

Irrigation system performance evaluation

Valsergio Barros da Silva
Janiquelle da Silva Rabelo
Antonio V. Rodrigues de Almeida

Irrigation system performance evaluation

Conventional sprinkling, drip and centre pivot

ScienciaScripts

Imprint
Any brand names and product names mentioned in this book are subject to trademark, brand or patent protection and are trademarks or registered trademarks of their respective holders. The use of brand names, product names, common names, trade names, product descriptions etc. even without a particular marking in this work is in no way to be construed to mean that such names may be regarded as unrestricted in respect of trademark and brand protection legislation and could thus be used by anyone.

Cover image: www.ingimage.com

This book is a translation from the original published under ISBN 978-3-330-99730-1.

Publisher:
Sciencia Scripts
is a trademark of
Dodo Books Indian Ocean Ltd. and OmniScriptum S.R.L publishing group

120 High Road, East Finchley, London, N2 9ED, United Kingdom
Str. Armeneasca 28/1, office 1, Chisinau MD-2012, Republic of Moldova, Europe
Managing Directors: Ieva Konstantinova, Victoria Ursu
info@omniscriptum.com

Printed at: see last page
ISBN: 978-620-8-61567-3

SUMMARY

GENERAL INTRODUCTION

Irrigation can be considered one of the most important agricultural practices for food production in the world, especially in regions where rainfall is irregular. In general, information on climatic and soil factors and the species to be grown is necessary when managing this cultural practice for vegetable production (COSTA, 2014). The amount of water to be applied to a given crop is related to the ability of the soil and vegetation to transfer water to the atmosphere (SILVA; RAO, 2006).

Irrigation in agriculture is a practice that can increase productivity and improve product quality, however, its use requires knowledge of factors that are decisive for irrigation management, which can directly interfere with the amount of water used, water storage in the soil, crop water requirements (SILVA, et al., 2010), and physiological aspects such as photosynthesis, leaf area and stomatal conductance (TAIZ; ZEIGER, 2013).

Distribution uniformity consists of the difference in the variation of water application across the whole plot, i.e. it depends on the variation of flow and service pressure across the whole irrigated plot. According to Bastos et al. (2009), if irrigation uniformity is low, it means that some areas receive less water and others receive more water than the required amount, causing damage to the normal development of the plants and leading to waterlogging, erosion, water loss through percolation and nutrient washing. According to Mantovani et al. (2009), although pressure control is a simple task, the lack of awareness of its importance means that most systems have a deficiency or total lack of infrastructure for pressure measurements, adjustments and control.

Determining the crop's water needs is one of the main premises for carrying out proper irrigation management and obtaining better crop quality and productivity. Furthermore, knowing that the crop is receiving the same amount of water throughout the growing area is fundamental for obtaining uniform stands and avoiding losses.

In the world, the greatest consumption of water occurs in agricultural activity (FAO, 2009; FERERES; SORIANO, 2007). According to the Food and Agriculture Organisation of the United Nations, approximately 10% of the water used in agriculture during the year would be enough to meet the water needs of 6.3 billion people during the same period. However, of every 100 litres of water used in agriculture, around 60 litres are lost through leaching or evaporation (FAO, 2009; FERERES; SORIANO, 2007).

According to Fereres and Soriano (2007), in more than half of irrigated agricultural crops the irrigation techniques used are of low efficiency, which generates excessive expenditure on water and electricity, in other words, low productive efficiency, which makes it necessary to optimise the efficiency of water use in irrigation. One of the existing alternatives for increasing the efficiency of water use in agriculture is through the adoption of irrigation systems that use little water but are able to put this in contact with the plants' root system (FERERES; SORIANO, 2007). Based on these observations, the aim of this study was to assess the distribution uniformity of drip, micro-sprinkler, conventional sprinkler and centre pivot irrigation systems.

REFERENCES

COSTA, R. A. **Okra crop subjected to drip irrigation as a function of evaporation in class a tank**. 2014. 44f. Thesis (Doctorate in Agronomy/Irrigation and Drainage) - Faculty of Agronomic Sciences, São Paulo State University, Botucatu-SP, February, 2014.

FERREIRA DF. 2010. SISVAR - **Statistical Analysis Systems**. Version 5.3. Lavras: Department of Exact Sciences, UFLA.

FERERES, E. AND M. A. SORIANO. Deficit irrigation for reducing agricultural water use. **J. Exp. Bot**. 58: 147-159, 2007.

FAO - Food and Agriculture Organisation of the United Nations. Available at: http://faostat.fao.org,Acesso on 06 Oct.2009.

MAROUELLI, W. A. & SILVA, W. L. C. Soil water tension limits for drip irrigated processing tomato cultivation. Brasília: **Embrapa Hortaliças**, 17p.Boletim de pesquisa e desenvolvimento / Embrapa Hortaliças, 37, 2008.

MAZOYER, M.; ROUDART, L. **The fallow farming systems of the temperate regions, The first agricultural revolution of modern times**.

IN: History of agriculture in the world: from the Neolithic to the contemporary crisis. São Paulo: Editora UNESP; Brasília, DF: NEAD, 2010. chap. 8, p.353-396.

SILVA, L. C.; RAO, T. V. R. Evaluation of methods for estimating peanut crop coefficients. **Brazilian Journal of Agricultural and Environmental Engineering**. Campina Grande. v. 10, n. 1, p. 128-131, 2006.

SILVA, V. G. F. Productive characteristics and water use efficiency in cotton plants under different irrigation strategies. **Brazilian Journal of Agricultural and Environmental Engineering**. Campina Grande. v. 14, n. 5, p. 451457, 2010.

TAIZ, L.; ZEIGER, E. **Plant Physiology**. 5. ed. Porto Alegre: Artmed, 2013. 918p.

CHAPTER 1

EVALUATION OF A PIVOT IRRIGATION SYSTEM INSTALLED AT FIELD LEVEL

ABSTRACT: The centre pivot irrigation method is being used more and more these days, as it is very efficient when properly sized. The aim of this study was to determine the efficiency of a centre pivot irrigation system located in the Jaguaribe-Apodi Irrigated Perimeter. The study was conducted in the experimental area of the Federal Institute of Ceará - IFCE / Campus Limoeiro do Norte. A centre pivot was evaluated, with the following characteristics: Manufacture, company Valmont Indústria e Comercio, centre pivot model VALLEY 4865 - 8000 - VL, 1 long throw 6.5/8" (55.11 m), with 25 m overhang and 26 m final cannon reach, 80.56 m pipe length, with a total irrigated radius of 106.56 m, equivalent to 3.57 ha, with 34 sprinklers with downpipes, 2,74 m above the ground, relief period at 100%, 2.62 hours/round, lamina applied per path of 2.56 mm, applying a flow per area of 9.76 m^3, pressure at the end of the pipe of 3.40 atm and pressure at the pivot point of 37.60 mca. The test to assess the uniformity of water distribution by the pivot was carried out by laying out two lines of collectors along the irrigation radius, perpendicular to each other. The collector lines were laid out from the pivot point to the end point, numbered in ascending order, with a distance of 5 metres between collectors. In this test, Fabrimar evaluation kits were used, consisting of collectors with a diameter of 8 cm, whose area is equivalent to 50.265 cm^2, suspended by aluminium rods. As the flow rate of the emitters in the centre pivot increases from the centre to the periphery, it is necessary to calculate the average volume of the blade collected by weighting the volumes measured by their position. This is done by multiplying each collected volume by its order number in the measured irrigated radius. The weighted average rainfall volume collected over the course of the test was 11.55mL per collection cup. The classification for the Christiansen Uniformity Coefficient, the pivot evaluated is in a good condition for water distribution. The centre pivot evaluated had regular CUC and poor CUD water distribution coefficients, which shows that the efficiency of the irrigation system used is in poor working order.

Keywords: Uniformity of application; irrigation; Christiansen Uniformity Coefficient

INTRODUCTION

Sprinkler irrigation has been used since the Second World War, with the invention of rotary sprinklers and the availability of reasonably cheap aluminium pipes. The first systems were very labour-intensive, because once the water had been applied to one row, the equipment had to be manipulated to irrigate the next row. Changes were made to improve the system, such as pumping the water through a main line where pipes were connected to form lateral lines that could be disconnected, moved and reassembled later. This had advantages in terms of reducing labour costs.

However, around 1950, Frank Zybach (Nebraska, USA) built the first centre pivot system. This system was patented in 1952 and mass-produced a year later.

Currently, the centre pivot irrigation method is being used more and more, as it is very efficient when properly sized. The aim of this study was to determine the efficiency of a centre pivot irrigation system located in the Jaguaribe-Apodi Irrigated Perimeter.

MATERIALS AND METHODS

The study was conducted in the experimental area of the Federal Institute of Ceará - IFCE / Campus Limoeiro do Norte. A centre pivot was evaluated, with the following characteristics: Manufacture, company Valmont Indústria e Comercio, centre pivot model VALLEY 4865 - 8000 - VL, 1 long throw 6.5/8" (55.11 m), with 25 m overhang and 26 m final cannon reach, 80.56 m pipe length, with a total irrigated radius of 106.56 m, equivalent to 3.57 ha, with 34 sprinklers with downpipes, 2.74 m above the ground, relief period at 100%, 2.62 hours/round, lamina applied per path of 2.56 mm, applying a flow per area of 9.76 m^3, pressure at the end of the pipe of 3.40 atm and pressure at the pivot point of 37.60 mca (figure 1).

Figure 1 - Centre pivot in the experimental area of IFCE / Campus Limoeiro do Norte

Source: Authors.

The test to assess the uniformity of water distribution by the pivot was carried out by laying out two lines of collectors along the irrigation radius, perpendicular to each other. The lines of collectors were laid out from the pivot point to the end, numbered in ascending order, with a distance between collectors of 5 m, as shown in illustration 2. Fabrimar's evaluation kits were used in this test, consisting of collectors (figure 3) with a diameter of 8 cm, whose area is equivalent to 50.265 cm^2, suspended by aluminium rods.

As the flow rate of the emitters in the centre pivot increases from the centre to the periphery, it is necessary to calculate the average volume of the blade collected by weighting the volumes measured by their position. This is done by multiplying each collected volume by its order number in the measured irrigated radius. Given by the following equation:

$$\bar{H} = \frac{\Sigma(Hi - Ni)}{\Sigma Ni} \quad \text{Eq. 1}$$

in which:

$\bar{H}$- weighted average blade of the collectors;

Hi - average blade collected at the i-th collector, ml;

N - number of collectors.

The Christiansen uniformity coefficient was determined using the following equation.

$$CUC = 100\left(1 - \frac{\sum_{i=1}^{N}(Hi - \bar{H})}{Ni\ \bar{H}}\right) \qquad \text{Eq. 2}$$

where: CUC - Christiansen's coefficient of uniformity, expressed as a percentage;

Hi - weighted blade collected at the i-th collector, ml;

$\bar{H}$ - weighted average blade of the collectors;

N - Number of collectors.

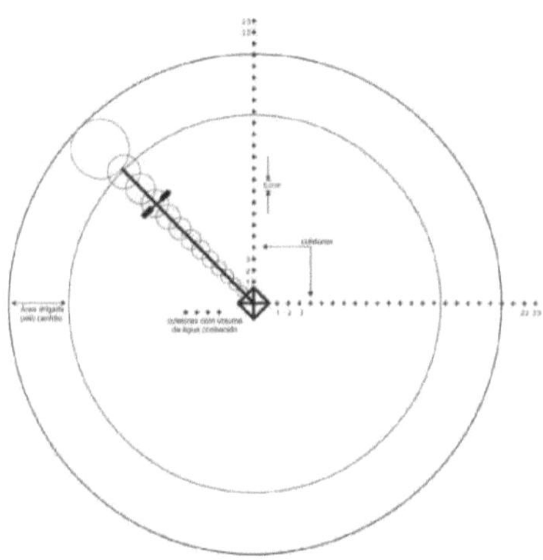

Figura 2 - arrangement of collectors in the field

Source: Authors.

Figura 3 - Fabrimar assessment collectors

Source: Authors

To determine the evaporation losses during the test, two collectors with a known volume of water were separated for each test. At the end of the test, the volume and the difference corresponding to the evaporation loss during the test were measured.

The coefficient of distribution uniformity (CUD) suggested by the US Soil Conservation Service was obtained using the equation:

$$CUD = \frac{h\ 25\%}{\bar{H}}\ x\ 100 \qquad \text{Eq. 3}$$

in which:

h 25% - average of 25% of the average slope of the collectors with the lowest rainfall.

During the test, the pressure at the pivot point was checked, the duration of the test, and the local wind speed was also measured using an anemometer.

RESULTS AND DISCUSSION

Water distribution efficiency along the pivot was assessed based on the following parameters:

Weighted average, Eq. 1.

$$\bar{H} = \frac{2667}{231} = 11{,}55\ ml$$

CUC: Christiansen's Uniformity Coefficient (%) is the coefficient that characterises the uniformity of water distribution, given according to Eq.2.

$$CUC = 100\left(1 - \frac{63{,}41}{21\ x\ 11{,}55}\right) = 73{,}85\%$$

CUD: Coefficient of distribution uniformity, which also characterises the uniformity of water distribution, given by Eq. 3.

$$CUD = \frac{8}{11{,}55}\ x\ 100 = 69{,}29\%$$

The values obtained in the field were calculated and organised according to the table below.

Table 1 - Values of the samples collected in the experimental area.

Order no. of collectors	Volume collected (mL)		Average volume (ml)	Weighted volume (Hix Ni)	(Hi-Hm)	Average volume/growth (ml)
	R1	R2				
1	20	20	20,00	20,00	8,45	0
2	20	10	15,00	30,00	3,45	6
3	15	14	14,50	43,50	2,95	11
4	15	13	14,00	56,00	2,45	11
5	16	17	16,50	82,50	4,95	12

6	15	13	14,00	84,00	2,45	**12,5**
7	12	12	12,00	84,00	0,45	**12,5**
8	10	12	11,00	88,00	0,55	**13**
9	12	14	13,00	117,00	1,45	**13**
10	15	10	12,50	125,00	0,95	**13,5**
11	16	12	14,00	154,00	2,45	**13,5**
12	12	10	11,00	132,00	0,55	**13,5**
13	15	12	13,50	175,50	1,95	**14**
14	15	12	13,50	189,00	1,95	**14**
15	14	16	15,00	225,00	3,45	**14**
16	14	16	15,00	240,00	3,45	**14,5**
17	12	14	13,00	221,00	1,45	**15**
18	12	15	13,50	243,00	1,95	**15**
19	12	13	12,50	237,50	0,95	**15**
20	12	0	6,00	120,00	5,55	**16,5**
21	0	0	0,00	0,00	11,55	**20**
231				**2667,00**	63,41	

Source: Authors

The volume of weighted average precipitation collected over the course of the test was 11.55mL per collection cup.

The graph below shows the water distribution profile along the pivot, where the x-axis plots the collector number and the y-axis plots the average blade. As the sprinklers move away from the pivot centre, the blade decreases and at the last collector the volume is the smallest. The classification for the Christiansen Uniformity Coefficient, according to the aforementioned author, shows that the pivot evaluated is in a good water distribution condition.

CONCLUSION

The central pivot evaluated had regular water distribution coefficients (CUC) and poor water distribution coefficients (CUD), which shows that the efficiency of the irrigation system used is in poor working order.

BIBLIOGRAPHICAL REFERENCES

ASAE - American Society of Agricultural Engineers. Field evaluation of microirrigation systems. St. Joseph, 1996. p.792-797 (CUD classification table).

BRALTS, V.F.; EDWARD, D.M.; WU, I.P. Drip irrigation design and evaluation based on statistical uniformity concept. In: HILLEL, D. (Ed). Advances in irrigation. Orlando: Academic Press, 1987. v.4, p.67- 117.

CHAPTER 2

DISTRIBUTION UNIFORMITY OF THE MICRO-SPRINKLER IRRIGATION SYSTEM IN THE CULTIVATION OF EARLY DWARF COCONUT TREES

ABSTRACT: The aim of this work was to assess the distribution uniformity of the micro-sprinkler irrigation system in the cultivation of early dwarf coconut trees. The experiment was carried out on a 3 ha area of early dwarf coconut trees, with 8 m x 8 m spacing, irrigated by micro-sprinklers. Four lateral lines were selected, which were located in the following positions: the beginning; 1/3 of the secondary line; 2/3 of the secondary line and the last.

After selecting the four lateral lines along the secondary, four points were selected along the lateral line in the following arrangements: first emitter, emitter at 1/3 of the length, emitter at 2/3 of the length and the last emitter.

The following parameters were assessed: coefficient of distribution uniformity (CUD), coefficient of absolute uniformity (CUa) and project application efficiency (Ea). The micro-sprinkler irrigation system proved to be suitable for irrigating coconut trees, with a good uniformity coefficient (86.67%) and acceptable application efficiency (91%). It is therefore capable of contributing to increased productivity and product quality in an efficient manner.

KEYWORDS: Cocos nucifera L. Water deficit. Efficiency.

INTRODUCTION

The coconut palm (Cocos nucifera L.) belongs to the Aracaceae family. It is a tall, erect palm, ranging in height from 10 to 20 m. Its fruit is used as a source of food, oil, fibre, fuel and other products (TAVARES, 2010).

The coconut crop requires a large amount of water during the vegetative growth and production phases. The ideal annual rainfall for the crop is approximately 1,500 mm, well distributed, with monthly values of more than 130 mm (MAHINDAPALA; PINTO, 1991).

Water deficit in the soil, caused by the poor distribution of rainfall throughout the year, is considered one of the main factors contributing to the low productivity of the coconut palm in the different regions where it is grown. Among the effects of water stress on the coconut palm are: a reduction in plant growth due to a decrease in the number of leaves and their size, premature leaf fall, a delay in the start of the production phase, a reduction in the number of female flowers per bunch, the fall of immature flowers and fruit and a reduction in the size of those that reach maturity (NAIR, 1989).

Irrigation can be considered one of the most important agricultural practices for food production in the world, especially in regions where rainfall is irregular. In general, information on climatic and soil factors and the species to be grown is necessary when managing this cultural practice for vegetable production (COSTA, 2014). The amount of water to be applied to a given crop is related to the ability of the soil and vegetation to transfer water to the atmosphere (SILVA; RAO, 2006).

Irrigation in agriculture is a practice that can increase productivity and improve product quality, however, its use requires knowledge of factors that are decisive for irrigation management, which can directly interfere with the amount of water used, water storage in the soil, crop water requirements (SILVA, et al., 2010), and physiological aspects such as photosynthesis, leaf area and stomatal conductance (TAIZ; ZEIGER, 2013).

Distribution uniformity consists of the difference in the variation of water application across the whole plot, i.e. it depends on the variation of flow and service pressure across the whole irrigated plot. According to Bastos et al. (2009), if irrigation uniformity is low, it means that some areas receive less water and others receive more water than the required amount, causing damage to the normal development of the plants and leading to waterlogging, erosion, water loss through percolation and nutrient washing. According to Mantovani et al. (2009), although pressure control is a simple task, the lack of awareness of its importance means that most systems have a deficiency or total lack of infrastructure for measuring, adjusting and controlling pressure.

Determining the crop's water needs is one of the main premises for carrying out proper irrigation management and obtaining better crop quality and productivity. In addition, knowing whether the crop is receiving the same amount of water throughout the growing area is fundamental for achieving homogeneity of the stand and avoiding losses.

Based on these observations, the aim of this study was to assess the distribution uniformity of the micro-sprinkler irrigation system when growing early dwarf coconut trees on a family farm in the Curu valley, Pentecoste - Ce.

MATERIAL AND METHODS

This work was carried out in the Curu-Pentecoste Irrigated Perimeter located in the Curu valley in the municipality of Pentecoste, Ceará, whose geographical coordinates are 03°44' South latitude 39°16' West longitude, 45 metres above sea level. The average annual rainfall is 797 mm. Rainfall is irregular and the average relative humidity is 67.6%. According to Alvares et al. (2014), the location has the BSh climate type, a semi-arid climate with low latitude and altitude.

The experiment was carried out in an area planted with 3 ha of early dwarf coconut trees, with a spacing of 8 m x 8 m, irrigated by micro-sprinkling, in a sub-unit of the 4 sectors of the property, which has 10 lateral lines, where each row of plants has a polyethylene lateral line 60 m long and 16 mm in diameter, with 6 emitters in each line and one emitter per plant with a nominal flow rate of 115 L h-1 with a spacing of 8 m between emitters. The secondary and main PVC lines are 50 mm in diameter.

Four lateral lines were selected on the bypass or secondary line in operation, which were in the following positions: beginning; 1/3 of the way along the secondary line; 2/3 of the way along the secondary line and last. After selecting the four lateral lines along the secondary line, four points were selected along the lateral line in the following positions: first emitter,

emitter at 1/3 of the length, emitter at 2/3 of the length and the last emitter (KELLER, KARMELLI, 1974).

Emitter flow measurements were taken at four emitters on each side, with three repetitions to obtain the average, with a volume of 1 litre for each collection, using a digital stopwatch and a 1000 ml measuring cylinder. The pressures were measured at the inlet and outlet of the lateral lines selected during the evaluation.

The following parameters were assessed: coefficient of distribution uniformity (CUD), coefficient of absolute uniformity (CUa), design application efficiency (Ea), average flow rate of all emitters (qa), average of the 25% lowest emitter flow rates (qn) and average of the highest emitter flow rates (qx). For the CUD and CUa data, the classification in Table 1 was taken into account.

The distribution uniformity coefficient (DUC) is based on the ratio between the minimum and average flow rates of the emitters in L h-1 (KELLER; KARMELI, 1974). The coefficient of absolute emission uniformity (CUa) is a modified form of the CUD equation proposed by Keller & Karmeli (1974), which includes the ratios of the maximum and minimum flow rates of the emitters to the average.

Table 1- Classification of CUD and Cua

CUA and CUa	Classification
90% or higher	Excellent
80 % a 90%	Good
70% a 80%	Regular
Less than 70%	Bad

Source: Bratls, (1986)

To calculate application efficiency, the following equation was used: Ea=Ks x CUD where: Ks is the transmissivity coefficient, which should be around 85% to 90% (FAO), for this experiment we used Ks=90%.

Table 2 - Classification of Ks for different soil types

Type of soil	Ks coefficient	Ks
Coarse sand	1,15	87
Sandy	1,00	91
Limoso	1,05	95
Clay	1,00	100

Source: Vermeiiren, (1997).

RESULTS AND DISCUSSION

The data was classified according to the Bralts (1986) and ASAE (1996) scales. The distribution uniformity coefficient (DUC) was 86.67% and was classified as good according to Bralts (1986) (Table 3).

In localised irrigation systems, according to López et al (1992), the CUD is the most widely used in evaluating irrigation system uniformity.

The results of this study were similar to those found by Saraiva et al. (2014), who obtained CUD values for twelve plots classified as good to excellent in micro-sprinkler irrigation for coconut cultivation. However, Santos and Resende (2014) obtained a CUD of 28.1% when evaluating a micro-sprinkler irrigation system for coconut cultivation.

The system showed an application efficiency (AE) of 91.00%

which, according to Bernardo (1995), is within the acceptable range for localised micro-sprinkler irrigation systems. The ideal and acceptable application efficiency (AE) for the different irrigation methods should be: ideal AE > 95% and acceptable AE > 80% (BERNARDO, 1995). Saraiva et al. (2014) found acceptable and some unacceptable application efficiency values for a total of 12 plots.

Table 3 - Average values of water distribution uniformity coefficients and classification of irrigation uniformity tests in coconut cultivation in the municipality of Pentecoste - Ce.

Variables assessed	Values obtained
V(L/h)	84,01
CUD (%)	86,67
CUa (%)	86,00
AS (%)	91,00
qa(L.h-1)	64,13
qn(L.h-1)	55,58
qx(L.h-1)	75,03

CONCLUSIONS

The micro-sprinkler irrigation system proved to be suitable for irrigating coconut trees, with a good uniformity coefficient and acceptable application efficiency. It is therefore capable of contributing to increased productivity and product quality in an efficient manner.

REFERENCES

ASAE - AMERICAN SOCIETY OF AGRICULTURAL ENGINEERS. Field Evaluation of Microirrigation Systems. St. Joseph, p.792-797. 1996.

BERNARDO, S. Manual de Irrigação. 6th ed. Viçosa: UFV, Impr. Univ., 1995. 657p.

BRALTS, V. F. Field performance and evaluation. In: NAKAYAMA, F. S.; BUCKS, D. A. (Ed.) Trickle irrigation for crop production. Amsterdam: Elsevier, 1986. p.216-240. (Development in Agricultural Engineering, 9).

COSTA, R. A. Okra crop subjected to drip irrigation as a function of evaporation in class a tank. 2014. 44f. Thesis (Doctorate in Agronomy/Irrigation and Drainage) - Faculty of Agronomic Sciences, São Paulo State University, Botucatu-SP, February, 2014.

KELLER, J.; KARMELI, D. Trickle irrigation design parameters. Transaction of the ASAE. St. Joseph, v.17, n.4, p.678-684, July/Aug., 1974.

LÓPEZ, J. R., ABREU, J. M. H.; REGALADO, A. P.; HERNÁNDEZ, J. F. G. Riego Localizado. Madrid, Spain: Mundi - Prensa, 1992. 405p.

MAHINDAPALA, R.; PINTO, J. L. J. G. Coconut cultivation. Lunuwila: Coconut Research Institute, 1991. 162p.

MANTOVANI, E. C.; BERNARDO, S.; PALARETTI, L. F. Irrigation: principles and methods. 3. ed. Viçosa: Editora UFV, 2009. 355p.

NAIR, R. R. Summer irrigation requirement of the coconut palm. Indian Coconut Journal, v.19, n.12, p.3-7, 1989.

SARAIVA, K. R.; REBOUÇAS, R. M.; SOUSA, F. Performance of a micro-sprinkler irrigation system in coconut cultivation. Revista Agropecuária Técnica. Fortaleza. v. 35, n. 1, p 62-68, 2014.

SANTOS, H. R. RESENDE, R. S. Micro-sprinkler Irrigation Uniformity under Dead Cover Conditions and Water Doses in the Dwarf Coconut Tree Crop. IV Embrapa Tabuleiros Costeiros Scientific Initiation and Postgraduate Seminar, 2014.

SILVA, L. C.; RAO, T. V. R. Evaluation of methods for estimating the coefficients of the
peanut crop. Brazilian Journal of Agricultural and Environmental Engineering. Campina Grande. v. 10, n. 1, p. 128-131, 2006.

SILVA, V. G. F. Productive characteristics and water use efficiency in cotton plants under different irrigation strategies. Brazilian Journal of Agricultural and Environmental Engineering. Campina Grande. v. 14, n. 5, p. 451-457, 2010.

TAVARES, M. F. F. Post-coco: adding value to the green coconut production chain. Available atwww.cspm.br/centraldecases , accessed on 21

November 2010.

TAIZ, L.; ZEIGER, E. Plant Physiology. 5. ed. Porto Alegre: Artmed, 2013. 918p.

CHAPTER 3

DRIP IRRIGATION SYSTEM PERFORMANCE IN CHERRY TOMATO CULTIVATION

ABSTRACT: A well-designed irrigation system allows water application uniformity of over 90%, which is considered a good index for irrigation. The aim of this study was to assess the performance of drip irrigation systems in terms of uniformity of water distribution and application to cherry tomatoes. The drip irrigation system was used for the study, with flow being collected from drippers on four lateral lines and four emitters on each of these (first, 1/3, 2/3 and last, totalling 16 emitters). Once the data had been obtained, the following parameters were evaluated: distribution uniformity coefficient, CUD, statistical uniformity coefficient, CUE, and the manufacturing variation coefficient - CVf. The classification of the coefficients was based on the values specified by ASAE (1994). The values obtained for the Distribution Uniformity Coefficients were CUD = 92.9%, CUE = 91.6% and CVf = 8.4%. The data obtained for CUD, CUE and CVf showed that the drip irrigation system performed well in terms of water distribution and application uniformity for the cherry tomato crop.

KEYWORDS: UNIFORMITY OF DISTRIBUTION; Solanum lycopersicum;

INTRODUCTION

Increased demand for water, mainly due to population growth and the resulting production activities, coupled with the environmental degradation of water bodies, has created a scenario of water scarcity in various regions of the world. This scarcity scenario has influenced changes in the population's habits, especially in irrigated agriculture, which has been seeking to reduce water consumption and optimise irrigation systems (SILVA, 2012).

Irrigation scheduling is very important when it comes to saving water in the soil. Determining the crop's water needs is one of the main premises for proper irrigation management (ZHANG et al., 2011).

A well-designed irrigation system can achieve water application uniformity of over 90%, which is considered a good index for irrigation. However, various factors can affect the uniformity of water distribution in localised irrigation systems, such as the emitter's working pressure, the speed of the water in the pipe, the alignment of the lateral line and the clogging of the emitters (SILVA, 2012).

Evaluating application uniformity is an essential activity in order to obtain the performance of drip irrigation systems (CUNHA et al., 2008).

Drip irrigation stands out for its efficiency and uniformity of water distribution, resulting in better utilisation of water resources and increasing the efficient use of water as well as crop productivity (SILVA, 2012). According to Souza et al. (2006), the water distribution uniformity coefficient and application efficiency are the main parameters used to evaluate drip irrigation systems, as they express the quality of irrigation and are decisive in the planning and operation of these systems.

Given the relevance of the subject, the aim was to assess the performance of a drip irrigation system in terms of water distribution and application uniformity in cherry tomato cultivation.

MATERIAL AND METHODS

The work was carried out at Prece (Cooperative Cell Education Programme), in the municipality of Pentecoste, state of Ceará, with geographical coordinates of 39°12'46" longitude and 03°55'20" latitude, and 56 m altitude. The experiment was carried out in an area of 570 m^2 growing *cherry* tomatoes *(Solanum lycopersicum* var. cerasiforme) using a drip irrigation system, with secondary lines 19 m long and lateral lines 29 m long.

For the evaluation, the katif brand self-compensating dripper was used, of the external type, spaced every 0.5 m, with a nominal flow rate of 3.75 L h^{-1} and a working pressure of 0.5 to 3.0 kgf cm^{-2} (Figure 1).

Figure 1: Katif dripper

A glycerine manometer was used to regulate the pressure to 2.0 kgf cm^{-2}, which was pressurised by a 3.0 hp pump. Flow measurements were taken from drippers on four lateral lines and four emitters (first, 1/3, 2/3 and last) on each of these, totalling 16 emitters, according to Keller & Karmeli (1975).

Water volumes were measured using a 1000 mL graduated beaker (figure 2). The collection time was set at 10 minutes, using chronometers accurate to 0.01s, and the flow rates were converted to litres per hour, the result of which was the average of three repetitions.

Once the data had been obtained, the following parameters were evaluated: distribution uniformity coefficient - DUC (Eq.1), statistical uniformity coefficient - SUC (Eq.2) and Manufacturing Variation

Coefficient - FVC .

Figure 2: Collection with graduated beaker

The distribution uniformity coefficient was determined using the methodology of Merrian and Keller (1978).

$$CUD = 100.\left(\frac{q25}{qmed}\right) \quad (1)$$

In which:

CUD - coefficient of distribution uniformity, %;

q25 - average flow of the lowest quartile, L h^{-1};

qmed - arithmetic mean of the collected flows, L h^{-1}.

$$CUE = 100.\left[1-\left(\frac{Sd}{qmed}\right)\right] \quad (2)$$

In which:

CUE - statistical coefficient of uniformity, %;

Sd - standard deviation;

qmed - arithmetic mean of the collected flows, L h^{-1}

$$CVf = \frac{Sd}{qmed} \quad (3)$$

Figure 3: Collectors

The interpretation of the coefficients was based on the values classified by ASAE (1994), as shown in Table 1.

Table 1 - Classification of coefficient of distribution uniformity (CUD), statistical coefficient of uniformity (CUE) and manufacturer's coefficient of variation (CVf) according to ASAE (1994)

Classification	CUD (%)	CUE (%)	Classification	CVf(%)
Excellent	100 - 94	100 - 95	Excellent	< 5
Good	87 - 81	90 - 85	Average	5 - 7
Reasonable	75 - 68	75 - 68	Marginal	7 - 11
Bad	62 - 56	62 - 56	Bad	11 - 15
Unacceptable	< 50	< 50	Unacceptable	> 15

RESULTS AND DISCUSSION

Table 2 shows the qmed values obtained for the dripper model analysed. You can see the measured values and the percentage differences between the flow rates supplied by the manufacturers (Nominal) and the values obtained in the field (Observed).

The observed variation in the average flow rate of the drippers was 6.7%, which is within the 7% limit accepted by the Brazilian standard NBR ISO 9261 (ABNT, 2006). It can be said that the drippers evaluated are within the specifications required by the technical standard.

Table 2 - Average flow (qmed) observed as a function of nominal service pressure.

Model	Katif
Nominal pressure (kgf cm^{-2})	2,0
qmed (L h^{-1}) Nominal	3,75
qmed (L h^{-1}) Observed	4,00
Variation (%)	6,7

Table 3 shows the values obtained for the uniformity coefficients of Distribution CUD = 92.9 % and Statistical CUE = 91.6%.

These values, according to ASAE (1994), are classified as good, showing that the drip system was well dimensioned for the local conditions and the crop.

Table 3. Classification of the coefficients of the drip irrigation system.

Parameters assessed	Values obtained (%)	Classification

CUD	92,9	Good
CUE	91,6	Good
CVf	8,4	Marginal

According to Bernardo et al. (2009), the Distribution Uniformity Coefficient (DUC) and Statistical Coefficient values above 90% are classified as excellent. The Manufacturing Coefficient was 8.4 % and according to ASAE (1994), it falls into the Marginal category.

It is worth noting that the tests on the irrigation system were carried out after the cherry tomato cycle and according to Resende et al, (2001) compromising the hydraulic characteristics of self-compensating drippers with the continuous use of water containing nutrients contributes to partial or total clogging of the emitters. According to Rodrigues et al, (2013) analysing uniformity coefficients is essential for assessing the performance of any irrigation system.

CONCLUSIONS

The data obtained from the evaluation coefficients CUD, CUE and CVf showed that the drip irrigation system performed well in terms of water distribution and application uniformity for cherry tomatoes.

REFERENCES

ABNT (BRAZILIAN ASSOCIATION OF TECHNICAL STANDARDS). NBR ISO 9261: agricultural irrigation equipment. Emitters and emitter tubes. Specification and test methods. São Paulo, 2006. 17p.

ASAE - American Society of Agricultural Engineers. Design and installation of microirrigation system. St. Joseph: ASAE, 1994. p.724-727.

BERNARDO, S.; SOARES, A. A.; MANTOVANI, C. Manual de irrigação, edition: 8ª. Publisher: UFV, 2ª reprint. Viçosa, MG, 2009, 625p.

CUNHA, F. F. et al. Micro-irrigation management based on system evaluation in melon cultivation. **Revista Caatinga**, Mossoró, v. 21, n. 3, p. 147-155, 2008.

Keller, J.; Karmeli, D. Trickle irrigation design. California: Rain Bird Sprinkler Manufacturing Corporation, 1975. 133p.

SILVA, L. P; SILVA, M. M. CORREA, M. M; SOUZA, F. C. D; SILVA, Ê. F. F. Performance of self-compensating drippers with different domestic sewage effluents. **Revista Brasileira de Engenharia Agrícola e Ambiental**, Campina Grande, PB, v.16, n.5, p.480-486, 2012.

RESENDE, R. S.; CASARINI, E.; FOLEGATTI, M. V.; COELHO, R. D. Occurrence of biological clogging in drip irrigation systems. Revista Brasileira de Engenharia Agrícola e Ambiental, v.5, p.156-160, 2001

RODRIGUES, R. R.; COLA, M. P. A.; NAZÁRIO, A. A.; AZEVEDO, J. M. G. DE; REIS, E. F. DOS. Efficiency and uniformity of a drip irrigation system in coffee growing. Ambiência Guarapuava (PR). 2013, 9, 2, 323-334

SOUZA, L. O. C.; MANTOVANI, E. C.; SOARES, A. A.; RAMOS, M. M.; FREITAS, P. S. L. Evaluation of drip irrigation systems used in coffee growing. Brazilian Journal of Agricultural and Environmental Engineering, v.10, p.541-548, 2006.

CHAPTER 4

EVALUATIONS OF THE HYDRAULIC PERFORMANCE OF A SPRINKLER IN AN IRRIGATED PASTURE AREA

INTRODUCTION

Irrigation is an agricultural technique that artificially applies water to plants using methods that best suit the soil and crop. According to Lima et al. (1999) irrigation can be defined as a set of techniques designed to move water in time or space in order to modify the agricultural possibilities of a region, with the aim of correcting the natural distribution of rainfall or supplying the real water needs of the crop, aiming for both high yields and products of high commercial value. This is an ancient technique, as literature reports that ancient civilisations have always settled on the banks of rivers, such as the Huang Ho and Iang-tsé-Kiang in China; the Nile in Egypt; the Tigris and Euphrates in Mesopotamia; and the Ganges in India (BERNARDO et al., 2006).

Sprinkler irrigation is the method in which water is sprayed over the surface of the ground, resembling rain, by fractionating the jet into a huge number of water droplets that spread through the air. This irrigation system is widely used due to the possibility of high distribution uniformity, adaptability to different crops and soils and easy control of the volume of water applied (ALENCAR et al., 2007). It is therefore important that new technologies are evaluated and disseminated in order to support the needs of an increasingly competitive agriculture, knowing that water is one of the main factors in crop development and the irregularity of the rainfall regime in some regions can become a restriction on agricultural development. With regard to the crop, the appropriate sprinkler type and height must be chosen (BERNARDO; SOARES; MANTOVANI, 2008).

Many factors can interfere with the level of water distribution uniformity, such as spacing between sprinkles, wind speed and direction, sprinkler operating pressure, sprinkler rotation speed and uniformity, among others (Christiansen, 1941). Sprinkler irrigation is greatly influenced by the action of the wind and various research studies have shown that an increase in wind speed causes a reduction in water distribution uniformity, as a result of the distortion of the water application profile and the reduction in the area covered by the sprinkler and, consequently, the reduction in the overlap of the water jets (Gomide et al., 1980; Ribeiro,1982; Vories & Bernuth, 1986; Paz, 1990; Martim-Benito et al.,1992; Alves & Castro, 1995).

As a consequence of the reduction in pasture productivity, livestock farming has been showing more critical levels of production over the years. One of the reasons for this is inadequate pasture management, especially with regard to the widespread practice of overstocking. For crops with medium-depth root systems, the distribution uniformity coefficient can vary between 70 and 80 per cent (FRIZZONE, 1992). The Christiansen

Uniformity Coefficient (CUC), developed in 1942, was the first index proposed and is the most widely used for quantifying water distribution uniformity in conventional sprinkler irrigation. There are several other coefficients proposed to assess the uniformity of water distribution from a lateral line, among which the most widely used, in addition to the Christiansen coefficient, is the coefficient of application uniformity (CUA). The aim of this study was to assess the performance of a sprinkler irrigation system in an irrigated pasture area at the Vale do Curu Experimental Farm (FEVC) located in the municipality of Pentecoste-CE by evaluating the Christiansen uniformity coefficient, application uniformity coefficient and potential application efficiency, evaporation loss and wind drag, lowest quartile application efficiency and lowest median application efficiency.

MATERIAL AND METHODS

The sprinkler irrigation system was evaluated in one of the pasture paddocks at the Vale do Curu Experimental Farm (FEVC) in the municipality of Pentecoste-CE. The following equipment was used to carry out the uniformity test: Fabrimar volumetric collectors for measuring rainfall, support, hoses or siphon, pitot tube for measuring pressures, clock for timing, measuring tape, graduated glass beaker for measuring the water collected, portable anemometer for measuring wind speed, bucket (container for collecting water).

A total of 80 collectors were used, and after distributing the collectors, the test time was set at 2 hours. Initially, a tape measure was used to demarcate the area of influence of the three sprinklers on the lateral lines observed, dividing it into square sub-areas of equal dimensions 3m x 3m. To measure the average flow of the emitters, plastic buckets were used to collect the water, a graduated cylinder (mm) was used to measure the water level and a clock was used to mark the time.

The final average flow rate of the emitters was obtained in Ls^{-1}, and then converted into m^3h^{-1}, which was determined by adding up the three repetitions that were measured in five seconds. The evaluation was carried out over a period of two hours and five minutes, which would ideally be equal to or half the irrigation time, given that the irrigation time is 11 hours, corresponding to the normal irrigation time of the pasture in the area. During the evaluation time, the wind speed was observed every 15 minutes using a portable digital anemometer at a height of 1.5 metres above the ground. Another parameter assessed during the test was the rotation of the sprinkler and the uniformity of rotation in its quadrants. The parameters used to evaluate the sprinkler system were:

Christiansen Uniformity Coefficient, distribution uniformity coefficient suggested by the US Soil Conservation Service, potential application efficiency, evaporation loss and wind drag can be determined, lowest quartile application efficiency and lowest median application efficiency.

RESULTS AND DISCUSSION

It was found that the sprinklers were operating at similar flow rates and applied an average flow rate of 1.4121 m^3h^{-1} to the soil, which may contribute to the application of a practically uniform average irrigation blade. This can be confirmed by the analysis of variance (Figure 1), which showed that there was no statistically significant difference between the flow rates applied by the different sprinklers.

COMPLETELY RANDOMISED EXPERIMENT ANALYSIS TABLE

FV	GL	SQ	QM	F
Treatments	2	0.00672	0.00336	0.0887 ns
Waste	6	0.22723	0.03787	
Total	8	0.23395		

** significant at 1% probability level (p < .01)
 * significant at 5% probability level (.01 =< p < .05)
ns not significant (p >= .05)

With regard to the uniformity of water distribution by the sprinklers, it was found that the Christiansen uniformity coefficient (CUC) and the distribution uniformity coefficient (CUD) showed relatively reasonable values for the type and system used, in the order of 79.12% and 66.82% respectively. These results corroborate the information provided by Olitta (1987), who states that in general the CUD values are lower than the CUC values.

According to Mantovani (2001), the uniformity coefficients mentioned above reached an acceptable value in the sprinkler irrigation system in the FEVC pasture area. They were classified as reasonable CUC and CUD, respectively.

Table 1. The results of the technical parameters determined when evaluating the performance of the sprinkler irrigation system in the FEVC pasture area are shown.

Table 1 - Technical parameters assessed in the pasture sprinkler system, Fortaleza CE, 2013.

CUC (%)	CUD (%)	EAp (%)	Pv (%)	EAq (%)	EAm(%)
79,12	66,82	73,92	26,08	49,39	58,48

According to Contrim et al. (1988) and IDE (1989), the difference between the results of the two coefficients (CUC and CUD) can be explained by the fact that the equation that determines the Coefficient of Uniformity of Distribution (CUD) only takes into account the average of 25% of the total amount of rainfall collected, and these 25% are usually in the first half of the lateral line, thus competing for the lowest weighting. The Christiansen Uniformity Coefficient (CUC), on the other hand, takes into account the average rainfall collected from all the rain gauges along the lateral, and is therefore statistically more representative.

It was observed that rainfall varied throughout the irrigated area, as the distance from the emitters decreased and rainfall was concentrated near the sprinklers, with points having higher rainfall and others lower, a fact that may have caused the low indices of the parameters analysed.

CONCLUSION

The results obtained in field conditions revealed that the system evaluated proved to be reasonable in terms of water application, with regard to the values of the coefficients CUC, CUD, EAp, Pv, EAq and EAm, since they were within the values recommended as acceptable by the literature, which means that the system in question is working with an acceptable degree of uniformity.

In view of the above, all the characteristics and physical agents that influence sprinkler irrigation are within acceptable limits according to the literature.

BIBLIOGRAPHICAL REFERENCES

ALENCAR, C.A.B. DE. et al. Analysis of automation in a fixed conventional minispray irrigation system. Revista de Engenharia na Agricultura, Viçosa, MG, v.15, n. 2, p. 109-118, Apr/Jun, 2007.

BERNARDO, S.; SOARES, A. A.; MANTOVANI, E. C. Manual de Irrigação. 8ª Ed. Viçosa. 2ª reprint: UFV, University Press, 2008. 625p.

BERNARDO, S.; SOARES, A. A.; MANTOVANI, E. C. Manual de Irrigação. 8th Ed. Viçosa: UFV, University Press, 2006. 625p.

CHRISTIANSEN, J.E. The uniformity of application of water bysprinkler systems. Agricultural Engineering, St Joseph, v.22,n.3, p.89-92, 1941.

FRIZZONE, J.A. **Irrigation by sprinklers**. Piracicaba: ESALQ - Department of Rural Engineering, 1992. 53p. Didactic Series, 3

LIMA, J. E. F. W.; FERREIRA, R. S. A.; CHRISTOFIDIS, D. Study on the use of water and electric power for irrigation in Brazil. In: II Water Meeting, 1999, Montevideo, Uruguay. Electronic annals... Available at: <http://www.iica.org.uy/p2-5.htm> Accessed on: 27 Nov. 2007.

MARTIN-BENITO, J.M.T.; GOMEZ, M.V.; PARDO, J.L. Working of sprinkler to optimise application of water. Journal of Irrigation and Drainage Engineering ASCE, New York, v.118,n.6, p.713-895, 1992.

SILVA, C.A.; SILVA, C.J. Evaluation of uniformity in localised irrigation systems. In: Revista Científica Eletrónica de Agronomia - ISSN 1677 - 0293, year IV, number 08, December 2005.

MIX
Papier aus verantwortungsvollen Quellen
Paper from responsible sources
FSC® C105338

Printed by Books on Demand GmbH, Norderstedt / Germany